Débora Perdigão Tejo
Maria Elvira Simionato
Carlos Henrique dos Santos Fernandes

Practical Manual for Growing Chives

Débora Perdigão Tejo
Maria Elvira Simionato
Carlos Henrique dos Santos Fernandes

Practical Manual for Growing Chives

Crop management for spring onions

ScienciaScripts

Imprint

Any brand names and product names mentioned in this book are subject to trademark, brand or patent protection and are trademarks or registered trademarks of their respective holders. The use of brand names, product names, common names, trade names, product descriptions etc. even without a particular marking in this work is in no way to be construed to mean that such names may be regarded as unrestricted in respect of trademark and brand protection legislation and could thus be used by anyone.

Cover image: www.ingimage.com

This book is a translation from the original published under ISBN 978-620-2-19610-9.

Publisher:
Sciencia Scripts
is a trademark of
Dodo Books Indian Ocean Ltd. and OmniScriptum S.R.L publishing group

120 High Road, East Finchley, London, N2 9ED, United Kingdom
Str. Armeneasca 28/1, office 1, Chisinau MD-2012, Republic of Moldova, Europe
Printed at: see last page
ISBN: 978-620-7-31617-5

SUMMARY

AUTHORS:

Débora Perdigao Tejo (Agronomy undergraduate)

North University of Paranà - UNOPAR

Londrina - Paranà

Maria Elvira Simionato (Agronomy undergraduate)

North University of Paranà - UNOPAR

Londrina - Paranà

Carlos Henrique dos Santos Fernandes (Agronomy undergraduate)

North University of Paranà - UNOPAR

Londrina - Paranà

Guilherme Lima Camargo (Agronomy undergraduate)

North University of Paranà - UNOPAR

Londrina - Paranà

Douglas Melo Lopes (Agronomy undergraduate student)

North University of Paranà - UNOPAR

Londrina - Paranà

Bruno Medeiros Kutlak (Agronomy undergraduate)

North University of Paranà - UNOPAR

Londrina - Paranà

Thais Cristina Morais Vidal (Agronomist, Master in Energy in Agriculture)

North University of Paranà - UNOPAR

Londrina - Paranà

Guilherme Renato Gomes (Agronomist, Master)

North University of Paranà - UNOPAR

Londrina - Paranà

Larissa Abgariani Colombo (Agronomist, PhD in Agronomy)

North University of Paranà - UNOPAR

Londrina - Paranà

We dedicate this work to our parents and family who have always encouraged us, supported us and given us all the support we needed in our studies.

ACKNOWLEDGMENTS

First of all, we thank God for guiding us along this path;

To Professor Thais, for her guidance and the opportunity to carry out this study;

To Mr. Valter, for all his help and dedication to the practical side of our work;

To the Universidade Norte do Paranà - UNOPAR, for providing the necessary space, supplies and materials.

1 PREFACE

The Practical Handbook for Growing Chives provides important information for planting and managing this species. It covers aspects such as planting time, recommended spacing, management of the main pests and diseases, as well as the symptoms they cause, climatic conditions, the main varieties, the main physiological disorders; proper soil management (fertilization and liming); irrigation; harvesting; post-harvest care, including packaging, storage and marketing of the final product.

Because it contains scientific information in easy-to-understand language, it can be used for technical purposes or even in the management of small and domestic productions. It also provides cost estimates (%) for each stage that the grower or future grower will invest in growing spring onions.

2 ORIGIN AND GEOGRAPHICAL DISTRIBUTION

The green onion or common onion (*Allium fistulosum*) is a species native to the East or Siberia. The species belongs to the Alliaceae family and is one of the most cultivated species by small farmers in all regions of Brazil (FILGUEIRA, 1982; MAKISHIMA, 1993).

This "green smell" is much appreciated by the population and is therefore widely grown in Brazilian homes (FILGUEIRA, 1982; MAKISHIMA, 1993).

Green onions are geographically distributed in practically all regions of the country, as they can withstand prolonged mild temperatures and there are cultivars with plants that tolerate heat well. There are few restrictions on its cultivation throughout the year and the ideal temperature range for cultivation is between 8 and 22°C, i.e. in mild conditions (COTIA, 1987; MAKISHIMA, 1993; FILGUEIRA, 1982).

3 ECONOMIC IMPORTANCE

The spring onion crop in both single and intercropping has a production capacity of 1.59 t ha^{-1} and an average of 19.70 t ha⁻ 1. These results show that the increase in production depends on the genotype and population used, depending on the carrying capacity of the environment and the production system adopted (BÜLL, 1993).

In 2016, there were approximately 525 hectares of spring onions cultivated throughout the state of Paranà, with a total production of 8,614 tons, according to SEAB - State Secretariat for Agriculture and Supply (2016). The area of spring onions harvested in Brazil in the 2006 harvest (the most recent data found in the literature) was 23,500 hectares, with a total production of 96,812 tons (IBGE, 2006).

In terms of the nutritional value of spring onions, every 100 grams contains: 19.5 kcal, 3.4 grams of carbohydrates, 1.9 grams of protein, 79.9 milligrams of calcium, 31.8 milligrams of vitamin C, 0.1 milligrams of pyridoxine B6, 26.9 milligrams of phosphorus and 1.6 milligrams of sodium (EMBRAPA, 2011).

4 SOIL AND CLIMATE REQUIREMENTS

4.1 CLIMATE

Temperature and photoperiod (day length) are climatic factors that influence the adaptation of spring onions and make it impossible to recommend the same cultivar for a wide range of latitudes. Cultivars not recommended for the wrong place and time result in low yields. As well as influencing production, temperature directly affects flowering (RESENDE; COSTA; SOUZA, 2016).

4.1.1 Temperature

Among the chive cultivars available on the market, there are those that can withstand prolonged cold, as well as cultivars that resist heat well, with few restrictions on planting them at any time of the year. The ideal temperature range for cultivation is between 8 and 22°C, i.e. in mild conditions (COTIA, 1987; MAKISHIMA, 1993; FILGUEIRA, 1982).

4.1.2 Precipitation

Excessive rainfall directly affects the yield of the spring onion crop and is absolutely related to the higher occurrence of leaf diseases and root pathogens (RESENDE; COSTA; SOUZA, 2016).

High relative humidity also causes damage related to leaf diseases, as well as increasing the cost of production and even compromising production (RESENDE; COSTA; SOUZA, 2016).

4.2 SOIL

The crop grows best in medium-textured soils with adequate levels of organic matter. Areas intended for growing spring onions should be free of compacted layers and have good drainage (RESENDE; COSTA; SOUZA, 2016).

To obtain a good yield, soil preparation is essential. The beds should follow the level curves of the land, thus avoiding water erosion, and should have a slight slope, so that water doesn't run off too quickly and water doesn't accumulate on the surface (RESENDE; COSTA; SOUZA, 2016).

4.3 WATER

It is an essential element for plant life and metabolism. Therefore, irrigation must be in accordance with the needs and tolerance of the species. Lack of water (water stress) increases or decreases the active ingredients depending on the cultivar used (PEREIRA; SANTOS, 2013).

Water has an effect on the speed of development and can affect the phytosanitary state and final quality of the crop (RESENDE; COSTA; SOUZA, 2016).

The amount of water required by the crop varies from 350 to 650 mm, depending on the following variables: climatic conditions, the cycle of the chosen cultivar and the irrigation system (RESENDE; COSTA; SOUZA, 2016).

5 CULTIVARS

Two species are cultivated: *Allium fistulosum* (green or common chives) and *Allium schoenoprasum* (thin-leaved or Galician chives). Varieties: Ano Todo (Figure 1), Evergreen Bunching, Fina Fresh, Fina Hosonegui, Fina Kurosenbon, Grossa Futonegui, Hanegui, Nebukai, Nirà, White Spear Bunching, White Tower (EMBRATER, 1980; COTIA, 1987; CAMARA, 1993; MAKISHIMA,1993; FILGUEIRA, 1982).

Figure 1. Chives of the Ano Todo variety.

Source: SIMIONATO, 2017.

6 SOIL ACIDITY AND LIMING

Chives are sensitive to soil acidity, growing best in soils with a pH of 6.0 to 6.5 and, based on soil analysis, the limestone applied should raise the percentage of base saturation to 70 or 80% (RESENDE; COSTA; SOUZA, 2016).

The lime should be applied 20 to 30 cm deep, as chives have a moderately deep root system (TRANI; CARRIJO, 2004).

The limestone should be incorporated at least 30 to 40 days before planting, preferably using finely ground limestone ("filler") with a PRNT of 80 to 90% or partially calcined (PRNT of 90 to 100%). The limestone should be applied 60 days before planting the crop if the limestone used is ordinary limestone (PRNT of 60 to 70%) (TRANI; CARRIJO, 2004).

7 MINERAL NUTRITION AND FERTILIZATION

Fertilization should be based on technical criteria, seeking high crop yields, but aspects related to commercial quality and post-harvest preservation should also be considered when making decisions about fertilization practices (FARIA; SILVA; MENDES, 2007).

7.1 PLANTING FERTILIZER

The whole process of sustainability, soil conservation and the environment must be taken into account when carrying out the practice of planting fertilizer on spring onions (TRANI, 2007).

It is necessary to check the nutritional status of the soil, and for this we use tools such as soil analyses, preferably up-to-date ones. Based on the data in this analysis, the amount of fertilizer/ha^{-1} is calculated and recommended. Nitrogen (N), phosphorus (P) and potassium (K) are the essential macronutrients most often used in planting fertilizers (MARSCHNER, 2012).

7.2 TOP DRESSING

The practice of top dressing on spring onions is used in order to maintain nutrients which, during the development of the crop, become low in the soil due to the plant retaining them in the soil solution (TRANI, 2007).

8 IMPLEMENTING CULTURE

The crop can be planted in two ways: directly in the seedbed using vegetative parts, or indirectly (Figure 2 - A) with seeds grown in trays (Figure 2 - B) (SOBREIRA FILHO, 2012; HENZ; ALCÂNTARA, 2009).

Figure 2. [A] Spring onion seedlings, in the early days of their development, in seedbeds. [B] Tray with substrate used for indirect sowing of spring onions.

Source: TEJO, 2017

7.3 PLANTING TIME

For successful production, it is advisable to grow in regions with temperatures of no more than 25°C, which are considered to have a temperate climate. In view of this temperature range, planting should be carried out during the periods of the year when the lowest temperatures are recorded, fall and/or winter (GONDIM, 2010).

In the South and Southeast, planting can be done all year round; in the North and Northeast, where the average annual temperatures are higher, it is recommended to plant between April and October and March and July, respectively. In the Center-West region, planting is recommended between April and August (GONDIM, 2010).

7.4 PLANTING SPACING

The most common spacing for spring onions is 15 or 20 cm between rows. The most common spacing between plants (pits) is 10, 15 or 30 cm, making it easier to manage, avoiding competition between plants and the spread of pathogens and pests. The appropriate spacing is shown in figure 3 (SOBREIRA FILHO, 2012; VAZ; JORGE, 2007).

Figure 3 - Diagram showing the spacing between rows and between plants.

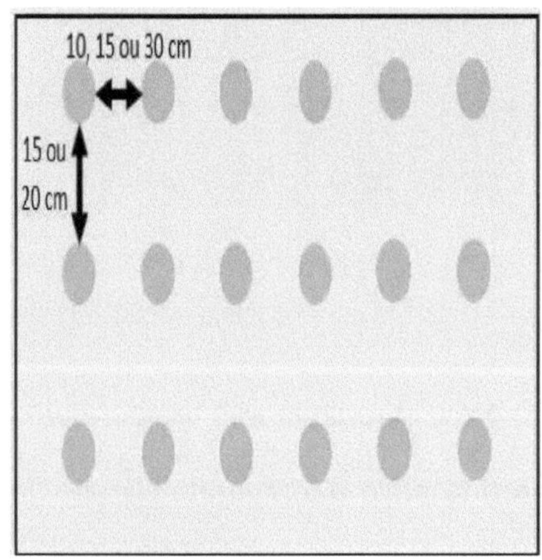

Source: TEJO, 2017.

7.5 TRANSPLANT

If seed propagation is used, germination should be checked between 7 and 15 days after sowing, and the seedlings should be transplanted into permanent beds between 30 and 40 days after sowing, when they should be

14

approximately 15 cm high (BOTELHO, 1987; SOBREIRA FILHO, 2012).

Pruning at the time of transplanting helps the seedlings to adapt better to the beds. The technique can be used simultaneously on both the aerial system (leaves) and the root system (roots) of the plants (ABREU; LIMA; MATTOS, 2014).

9 IRRIGATION

Irrigation is essential when planting spring onions, as this species grows quickly and produces a large amount of green mass (PEREIRA; SANTOS, 2013). Watering is done daily and the water used must be clean and of good quality (MAKISHIMA, 1993; VAZ; JORGE, 2007).

9.1 Sprinkler irrigation

Conventional sprinkler irrigation is the most widely used method, due to its adaptability to leafy vegetables, ease of management and adequate water supply. The system should be installed as soon as the beds are raised, with a spacing of 12 x 12 m between sprinklers (SIMOES *et al.*, 2011).

9.2 Drip irrigation

The drip irrigation system should be installed after the beds have been built. Two drip tubes are installed per bed, spaced no more than 30 centimeters apart. After installation, the system should be turned on and tested for possible leaks (SIMOES *et al.,* 2011).

10 PEST CONTROL

In order to manage pests properly, it is important to know which insects attack the spring onion crop, as well as their habits, time of occurrence and damage (COSTA, 2002).

10.1 Rose caterpillar (*Agrotis ipsilon*) (Lepidoptera: Noctuidae):

They are moths with brown forewings with black spots and white hindwings with grayish edges (Figure 4 - A). This insect has a high biotic potential and a female can lay up to 1,260 eggs during her lifetime. The young form of this pest are robust, gray-brown caterpillars. Due to their habit of curling up, they are popularly known as pink caterpillars (Figure 4 - B) (COSTA, 2002).

Figure 4. [A] Adult of *Agrotis ipsilon*. [B] Young form (caterpillar).

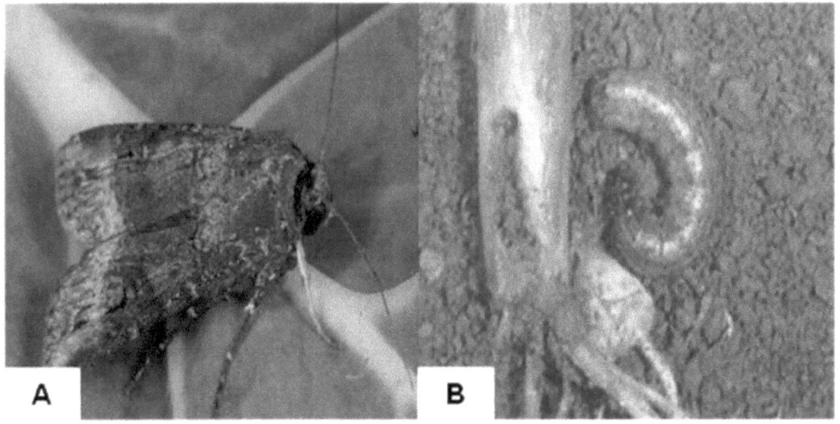

Source: QUINTELA, 2007.

The damage caused by the rosca caterpillar is the cutting of plants at the height of the plant's neck, causing a reduction in the stand. The control of the pest involves the destruction of crop remains, disturbance and exposure to the sun when preparing the area, as well as proper fertilization and irrigation, within the recommended levels. For chemical control, the application of products such as carbaryl, directed at the neck of the plants, is recommended (COSTA, 2002).

10.2 *Thrips* (*Thrips tabaci*) (Thysanoptera: Thripidae):

The thrips or louse is considered one of the most important pests of the spring onion crop (Figure 7). The adult has an elongated body and narrow, fringed wings. It measures approximately 1 mm in length and is light yellow or brown in color (Figure 8). The nymphs are greenish-yellow and have no wings (COSTA, 2002).

The damage is caused by white spots that turn silver. When the attack is intense, the plants become twisted, the leaves turn yellow and dry out (COSTA, 2002).

The most widely used control method for thrips in spring onions is chemical, using phosphorus and pyrethroid insecticides (COSTA, 2002).

Figure 5. [A] Damage caused by *Thrips tabaci on spring* onion leaves. [B] Adult form.

Source: OLIVEIRA, 2009.

10.3 Miner fly (*Liriomyza trifolii*) (Diptera: Agromyzidae):

The adult of this pest is a fly approximately 1mm long, black in color with yellow spots on the head and between the wings (Figure 6 - A). They make mines or galleries in the leaves, reducing the plant's ability to carry out photosynthesis (Figure 6 - B).

Not planting crops that are susceptible to the miner fly in the vicinity of the area used to grow spring onions and destroying crop remains are very important cultural measures for controlling the miner bug (COSTA, 2002).

For chemical control, it is recommended to apply products (insecticides) registered for the crop (COSTA, 2002).

Figure 6. [A] Pupae and larva. [B] Symptoms of the leaf miner on a spring onion leaf.

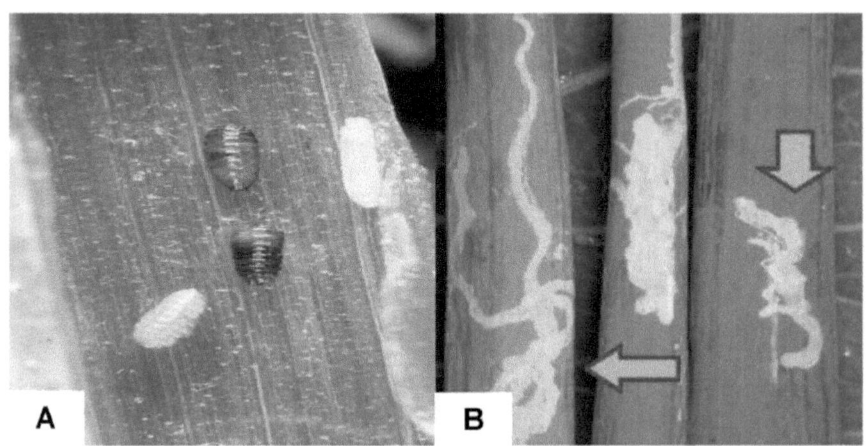

Source: Portal de Conteùdo Agropecuârio, 2006.

11 DISEASE CONTROL

During the production process, spring onions are susceptible to various diseases, with their degree of importance varying according to the growing conditions (PEREIRA; OLIVEIRA; PINHEIRO, 2014).

11.1 Mildew (*Peronospora destructor*):

Also known as black wool, this disease usually occurs in conditions of low temperature and high humidity. Symptoms begin on older leaves, with elliptical lesions covered in dark gray mycelium. As it progresses, it can affect the entire leaf, causing yellowing, folding of the leaves and death, as can be seen in figures 7 - A and 7 - B (PEREIRA; OLIVEIRA; PINHEIRO, 2014).

Figure 7. [A] *Peronospora destructor* pathogen. [B] Severe mildew symptoms on onion leaves.

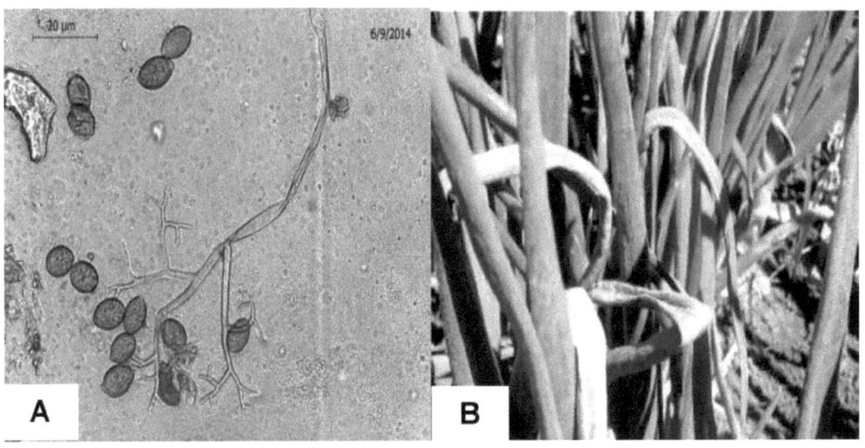

Source: Plant Pathology, 2014.OLIVEIRA, 2014.

Control can be preventative, using healthy, good quality seedlings. The soil should be well-drained and excessive nitrogen fertilization is not recommended. Do not plant in high densities and avoid growing chives next to other alliums. Fungicides with active ingredients such as chlorothalonil, propineb, cymoxanil, folpet, cyazofamid, mancozeb, among others, are used for chemical control (PEREIRA; OLIVEIRA; PINHEIRO, 2014).

11.2 Tip blight (*Botrytis squamosa*):

Several factors can cause leaf burn in spring onions, making it difficult to diagnose this abnormality. Some pathogens cause secondary infections in the lesions of *Botrytis squamosa* (Figure 8 - A), making accurate diagnosis difficult. The symptoms presented are 1 to 3 mm whitish spots on the tips of the leaves, as shown in Figure 8 - B (PEREIRA; OLIVEIRA; PINHEIRO, 2014).

Figure 8. *Botrytis squamosa* pathogen. *Botrytis squamosa* symptoms on spring onion leaves.

Source: APS, 2003. CARVALHO, 2014.

As control measures, you can adopt crop rotation and not plant other alliums in the area for at least two years; avoid high planting densities; if the irrigation

22

method is sprinkler irrigation, do it in the morning, making it easier for the leaves to dry. Fungicides based on mancozeb and captan are used to control the disease (PEREIRA; OLIVEIRA; PINHEIRO, 2014).

11.3 Rust (*Puccinia allii*):

The right conditions for the crop's development are low rainfall and high relative humidity. Its severity varies according to the stage of development the plant is at and the climatic conditions. The symptoms are small, whitish dots on the leaves, which become orange, circular pustules, about 1 to 3 mm long (Figure 9 - A) (PEREIRA; OLIVEIRA; PINHEIRO, 2014).

As time goes by, these spots start to expose a yellowish powdery mass, which is the fungus's urediniospores (spores) (Figure 9 - B). Leaves with high severity turn yellow and dry out (PEREIRA; OLIVEIRA; PINHEIRO, 2014).

As control measures for the disease, planting in lowland soils and soils with compacted layers should be avoided; elimination of the remaining plants; balanced fertilizations, with attention to nitrogen; fungicides based on mancozeb and copper oxychloride are efficient (PEREIRA; OLIVEIRA; PINHEIRO, 2014).

Figure 9. [A] Symptoms of rust caused by *Puccinia porri* on spring onions. [B] *Puccinia allii* pathogen.

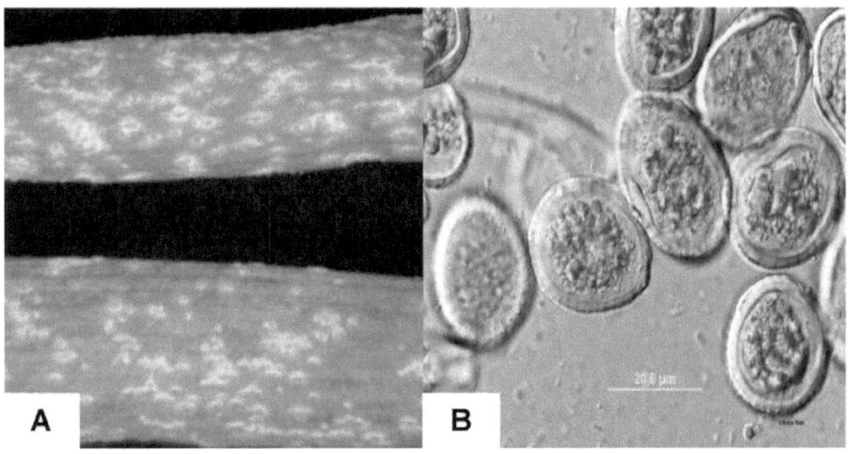

Source: OLIVEIRA, 2014. THE UNIVERSITY OF MAINE, 2010.

11.4 Purple spot (*Alternaria porri*):

This disease is caused by the fungus *Alternaria porri* (Figure 10 - A), which is common in hot and humid climates. The initial symptoms are watery, elliptical-shaped lesions on the leaves. These lesions then become larger and take on a straw color with a purple tinge in the center. In the center of these lesions, there is a grayish color, which is the fruiting of the pathogen (Figures 10 - B). As the disease progresses, the lesions coalesce, the leaves wither and wrinkle from the apex (PEREIRA; OLIVEIRA; PINHEIRO, 2014).

Figure 10. [A] *Alternaria porri* pathogen. [B] Initial lesions of the disease.

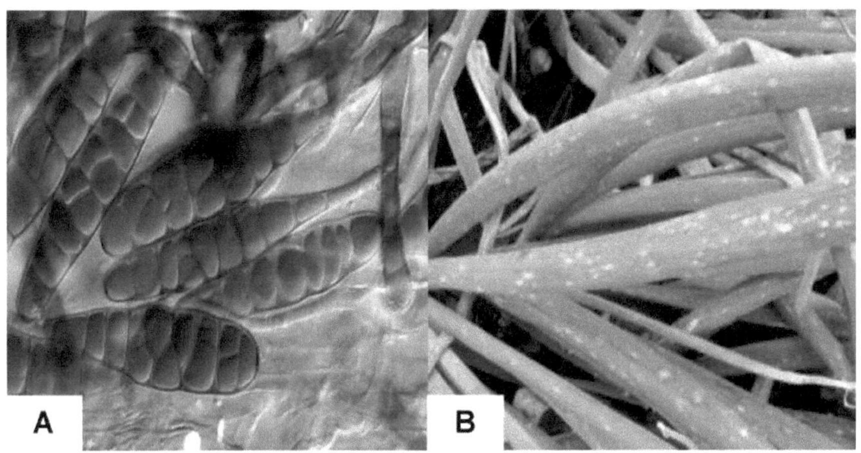

Source: Fungi of Great Britain and Ireland, 2009. OLIVEIRA, 2014.

Control can be preventative, with practices that include the correct choice of planting area, without the accumulation of humidity and strong winds, use of healthy seeds and seedlings, crop rotation, incorporation of cultural remains, lower population densities, planting in drained soils, elimination of alternative hosts and sprinkler irrigation (PEREIRA; OLIVEIRA; PINHEIRO, 2014).

Fungicides such as copper oxychloride, mancozeb, tebuconazole, metiram, iprodione, tebuconazole, azoxystrobin, prochloraz and others can be used for chemical control. These products are applied through spraying, which should be carried out as soon as the plants show the first symptoms (PEREIRA; OLIVEIRA; PINHEIRO, 2014).

12 PHYSIOLOGICAL DISORDERS

12.1 DISORDERS CAUSED BY NITROGEN DEFICIENCY (N)

Nitrogen participates in cell structures, such as enzymes and coenzymes, amino acids, vitamins, chlorophyll and proteins, and is directly linked to absorption and photosynthesis processes (MENDES *et al.*, 2008).

A deficiency in chives causes a slowdown in growth and a fasciculated root system. Older leaves turn a light green color and turn yellow-green, reaching a light straw color at the tips of the leaves and younger leaves show dwarfism and become delicate (MENDES *et al.*, 2008).

12.2 DISORDERS CAUSED BY POTASSIUM (K) DEFICIENCY

Potassium is present in protein synthesis and osmotic processes, and is also responsible for coordinating pH control and the opening and closing of stomata (MENDES *et al.*, 2008).

The disorders caused by potassium deficiency show the following symptoms: the ends of the older leaves show necrosis with a light straw color that progresses to a darker shade, and the plant's overall development is reduced. This makes it easier for pests and fungi to attack (BELFORT; HAAG, 1983).

12.3 DISORDERS CAUSED BY PHOSPHORUS (P) DEFICIENCY

The phosphorus required by spring onions is lower than other nutrients such as nitrogen and potassium, but it is still considered a key nutrient for good production. Phosphorus is part of carbohydrate esters, coenzymes, phospholipids and nucleic acid, taking part in symbiotic N fixation and energy storage (MENDES *et al.*, 2008).

The disturbances caused by phosphorus deficiency are old leaves turning yellow and then drying out easily and the intermediate and new leaves turning a dark green color (RESENDE; COSTA; YURI, 2015).

13 CROP ROTATION AND INTERCROPPING

13.1 CROP ROTATION

Crop rotation is indicated in medicinal plant production areas, by interspersing species from different botanical families over time (PEREIRA; SANTOS, 2013). Species from the Poaceae family are the most suitable for crop rotation with chives, especially corn and millet (HASSE; MAY-DE MIL; LIMA NETO, 2007).

Medicinal and aromatic plants, including chives, parsley, mint and burdock, in crop rotations have been shown to be effective in reducing both the incidence and severity of cruciferous hernia, which is the most prominent pathology in the brassica genus (HASSE; MAY-DE MIL; LIMA NETO, 2007).

13.2 CONSORTIZATION

The adoption of intercropping in vegetables, in addition to increasing final production and therefore profits, brings advantages such as more effective use of the soil, nutrients and other resources made available to the plants (SILVA; HEREDIA, 1983; SULLIVAN, 1998).

It is advisable to intercrop with plants that have the same climatic characteristics as spring onions, such as African spinach and New Zealand spinach, which adapt easily to tropical climates (FILGUEIRA, 1982).

The intercropping of these species is indicated if the aim is to benefit the spring onion crop and increase its dry mass productivity, as this association is not effective when the interest is in the spinach crop (ZARATE; VIEIRA, 2004).

14 HARVESTING, PACKING AND GRADING

14.1 HARVESTING

Harvesting of the leaves begins between 55 and 60 days after planting or between 85 and 100 days after sowing, when the leaves reach a height of 0.20 to 0.40m. The leaves should be harvested whole, close to the base. The whole plant can also be harvested to make the most of the pseudostem. The leaves are cut between 10 and 15 cm from the ground level or above the apical bud. The regrowth is used to make new cuts, allowing the crop to be exploited for two to three years, in which case harvests are carried out every 50 days, especially when managed in mild climate conditions (EMBRATER, 1980; COTIA, 1987; MAKISHIMA, 1993; FILGUEIRA, 1982).

14.2 PACKAGING

Once dried, the green onions should be packed and stored in plastic bags, glass or containers that reduce the absorption of moisture by the plant. The packaging should not vary in color, so that there is no negative interference with the leaves in relation to the incidence of light, and it should be made of a material that does not spread odors, so that there is no contamination of the marketed product, in this case spring onions. It is worth noting that good quality packaging extends the shelf life of the cultivar (SHEPHERD, 1993).

The leaves of spring onions should be stored in airtight packaging, labeled with the name of the species and the date of harvest, in an airy, dark and dry place. Among the functions of packaging are transportation, sale, aspects related to appearance, identification, economic visibility, as well as preventing mechanical damage (SHEPHERD, 1993).

The crate is a widely used packaging for leafy vegetables, but its wide gaps cause a lot of damage to the leaves and it doesn't protect against wind and sunlight. Grass is used to protect these products (VADA, 1999).

14.3 CLASSIFICATION

Green onions are classified by the color of their leaves, which is one of the first factors to be considered by the final consumer and can be a determining factor in the purchase of the product. It has a strong influence on quality and is indicative of the stage of ripeness when the leaves are straw-colored. It has a strong correlation with nutritional function, being associated with chlorophyll levels, and is more nutritious when it is in an optimal state of preservation (CUI; XU; SUN, 2004; KASIM; ERKAL, 2008).

15 MARKETING

The *Allium spp.* spice plant, popularly known as chives, has particular characteristics in various parameters, even in the market where it is sold (PEREIRA; SANTOS, 2013).

The consumer market always demands good products, so the quality of the end product is of paramount importance. For this to happen, the handling and processing of the crop while it is still in the field must be taken into account, as well as the availability of labor, environmental conditions, drainage and access routes (PEREIRA; SANTOS, 2013).

With these characteristics, it is possible to identify the species that are best suited to the available conditions, always bearing in mind the lowest production costs among the criteria of good manufacturing practices, as well as having as a central objective the production of good quality products that please the end consumer (PEREIRA; SANTOS, 2013).

Some suggestions that producers can follow are: researching the market and analyzing the quality of the product on offer; acquiring as much technical information as possible about the species selected; at the beginning of production, allocating a smaller area to the cultivation of spring onions, in order to verify the viability of production in the area, as well as checking possible climatic influences on the cultivation of this crop in the region; contacting buyers, agreeing on purchase contracts and training the workforce (FURLAN, 1998).

There are various ways of earning income from the sale of spices, such as small-scale production in seedling nurseries, where the final product is destined directly for restaurants and handicrafts made with spices, and large-scale production for wholesalers or companies that sell bagged spices (PEREIRA; SANTOS, 2013).

16 PRODUCTION COSTS

After obtaining the information about the crop, the grower should make a spreadsheet as shown in Table 1, including all the inputs and services, in order to know the total costs of planting the crop (SIMOES *et al.*, 2011).

TABLE 1 - Estimated cost of production (%) of spring onions on 1 ha/year.

INPUTS AND SERVICES	UNIT*	QUANTITY	COST (IN %)
Mineral fertilizer (04 - 14 -08)	t	2	9,57
Mineral fertilizer (Borax)	Kg	20	0,23
Mineral Fertilizer (Sulfate of Ammonia)	t	0,5	2,4
Mineral Fertilizer (Zinc Sulfate)	Kg	20	0,22
Organic Fertilizer (Manure)	t	10	5,82
Energy Power for irrigation	kwh	1733	2,53
Seeds	Kg	2	4,5
Substrate for seedlings	Sc	45	3,55
Fertilization (Manual of Coverage)	d/h	6	1,17
Fertilizers (With tractor)	h/m	4	1,55
Weeding (Manual)	d/h	90	17,45
Harvest and post-harvest	d/h	200	38,79

Irrigation (Sprinklers)	d/h	8	1,55
Irrigation (Assembly of system)	d/h	2	0,38
Seedlings (Formation in trays)	d/h	15	2,9
Soil preparation (plowing)	h/m	3	1,17
Soil preparation (harrowing)	h/m	3	1,17
Soil preparation	h/t	4	1,55
Transplanting	d/h	18	3,5

* Unit: t= ton; kwh= kilowhatts hour; Kg= kilogram; Sc= bag; d/h= day/man; h/t= hour/tractor; h/m= hour/machine.

Source: Adapted from EMATER-DF (2016).

Among the production costs, harvesting and post-harvesting are the ones that stand out, requiring the most investment (38.79%).

REFERENCES

ABREU, P. T.; LIMA, M. A. C.; MATTOS, J. D. A. Influence of Seedling Preparation on the Production of *Allium Fistulosum*. *Horticultura Brasileira*, v. 22, n. 2, p. 441. 2004

APS. *Botrytis squamosa* pathogen. Available at: < https://www.apsnet.org/publications/imageresources/Pages/IW000071.aspx>. Accessed on: September 20, 2017.

BELFORT, C. C.; HAAG, H. P. Nutriçâo mineral de hortaliças: LVI-carência de macro nutrientes em cebolinha (*Allium schoenoprasum*). *Anais da Escola Superior de Agricultura Luiz de Queiroz*, v. 40, n. 1, p. 221-234. 1983.

BOTELHO, M. Cultivo de Ervas Aromâticas e Temperos. Rio de Janeiro: Editora Tecnoprint, p. 149. 1987.

BÜLL, L. T. Corn cultivation: factors affecting productivity. Piracicaba: POTAFOS, p. 301. 1993.

CÂMARA, F. L. A. Nutrition and fertilization of mandioquinha-salsa, parsley, coriander, leeks, spring onions and peppers. In: SYMPOSIUM ON NUTRITION AND FERTILIZATION OF VEGETABLES, Jaboticabal. Piracicaba: POTAFOS, 1993. p. 473-476. 1990.

CARVALHO, A. D. F. Botrytis sp. symptoms on onion leaves. IN: PEREIRA, R.B.; OLIVEIRA, V. R.; PINHEIRO, J. B. Diagnosis and management of fungal diseases in the onion crop. *BRASILIA: Embrapa (Technical Circular 133)*, p. 2-16. 2014.

COTIA AGRICULTURAL COOPERATIVE. Manual for growing the main vegetables. Cotia: Central Cooperative - Department of Seeds and Seedlings-DIA. 1987.

COSTA, N. D. The Onion Crop. Brasilia: Embrapa Informaçâo Tecnológica (Coleçâo Plantar: 45), p. 59-72. 2002.

CUI, Z. W.; XU, S. Y.; SUN, D. W. Effect of microwavevacuum drying on the carotenoids retention of carrot slices and chlorophyll retention of Chinese chive leaves. Drying Technology, Philadelphia, v. 22, n. 3, p. 563-575. 2004.

RESENDE, G. M.; COSTA, N. D.; YURI, J. E. Growing spring onions: cultivars and adjustments to NPK fertilizer recommendations for the Sâo Francisco Valley. *Embrapa Semiàrido-Technical Communication (INFOTECA-E)*. 2015.

EMATER - DF. Brazilian Technical Assistance and Rural Extension Company. Production Costs. Available at: <http://www.emater.df.gov.br>. Accessed on: September 25, 2017.

EMATER. Brazilian Technical Assistance and Rural Extension Company. Olericulture Technical Manual. Rio de Janeiro: Brasilia, p. 98 (Manuals, 28). 1980.

EMBRAPA, 2011. Brazilian Agricultural Research Corporation. Table of Nutritional Composition of Vegetables. Available at:

<https://www.embrapa.br/documents/1355126/9124396/Tabela%2BNutricion al%2Bde%2BHortali%25C3%25A7as/d4ae0965-9e94-4f19-a20e-b7721bdc1266>. Accessed on: 18 Nov. 2017.

FARIA, C. M. B.; SILVA, D. J.; MENDA. M. S. Onion cultivation in the Northeast. Embrapa Semi-Arido. Sistemas de Produçâo, 3 (Electronic Version). 2007.

FILGUEIRA, F.A.R. Manual de olericultura: agrotecnologia moderna na produçâo e comercializâo de hortaliças. Viçosa: UFV.1982.

Fungi of Great Britain and Ireland. *Alternaria porri*. Available at: < http://fungi.myspecies.info/all-fungi/alternaria-porri>. Accessed on: September 20, 2017.

FURLAN, M. R. Herbs and spices: cultivation and marketing. In: PEREIRA, R. C.; SANTOS, O. G. Plantas Condimentares: Cultivo e Utilizaçâo. Fortaleza: *Embrapa Agroindustrial Tropical (Documentos - 161)*, p. 42. 2013.

GONDIM, A. Catàlogo brasileiro de hortaliças. *Brasilia: Embrapa*, p. 24. 2010.

HASSE, I.; MAY DE MIO, L. L.; LIMA NETO, V. C. Effect of pre-planting with medicinal and aromatic plants on the control of *Plasmodiophora brassicae*.

Summa Phytopathologica, v. *33*, n. 1, p. 74-79. 2007.

HENZ, G. P.; DE ALCÂNTARA, F. A. Vegetable gardens: the producer asks, Embrapa answers. *Headquarters Information Area-Colec Criar, Plantar, ABC, 500P/500R (INFOTECA-E)*. 2009.

IBGE, 2006. Brazilian Institute of Geography and Statistics. Table 1706 - Production, sales and value of horticultural production in agricultural establishments. Available at:

<https://sidra.ibge.gov.br/tabela/1706#resultado>. Accessed on: 06 Nov. 2017.

KASIM, M. U.; KASIM, R.; ERKAL, S. UV-C treatments on fresh-cut green onions enhanced antioxidant activity, maintained green color and controlled 'telescoping'. Journal of Food Agriculture & Environment, Finland, v. 6, n. 3-4, p. 63-67. 2008.

MAKISHIMA, N. The cultivation of vegetables. *Brasilia: EMBRAPA-CNPH: EMBRAPA-SPI (Coleçao plantar, 4)*, p. 116. 1993.

Marschner P. Mineral nutrition of higher plants. Australia, Academic Press. ed. 3ª , p. 651. 2012.

MENDES, A. M. S.; FARIA, C. M. B.; SILVA, D. J.; RESENDE, G. M.; OLIVEIRA NETO, M.B.; SILVA, M. S. L. Nutriçâo mineral e aduçâo da cultura da cebolinha no Submédio do Vale do Sâo Francisco. *Embrapa Semiàrido-*

Technical Circular (INFOTECA-E). 2008.

OLIVEIRA, V. R. Thrips. Available at:

<http://www.agencia.cnptia.embrapa.br/gestor/cebola/arvore/CONT000gnn6ir oc02wx5ok0cdjvsccyctk5v.html>. Accessed September 13, 2017.

PEREIRA, R. C.; SANTOS, O. G. Condiment Plants: Cultivation and Use. Fortaleza: *Embrapa Agroindustrial Tropical (Documentos - 161),* p. 42. 2013.

PEREIRA, R.B.; OLIVEIRA, V. R.; PINHEIRO, J. B. Diagnosis and management of fungal diseases in onion crops. *BRASILIA: Embrapa (Technical Circular 133),* p. 2-16. 2014.

PLANT PATHOLOGY. Mildew pathogen. Available at: <

https://plantpathology.ces.ncsu.edu/2014/06/cucurbit-downy-mildew-outbreak-in-southeastern-north-carolina/>. Accessed September 20, 2017.

PORTAL DE CONTEÙDO AGROPECUARIO. Available at: <https://www.agrolink.com.br/culturas/problema/larva-minadora_383.html>.

Accessed on: September 13, 2017.

QUINTELA, E. D. Onion cultivation in the Northeast. Available at: <http://www.cpatsa.embrapa.br:8080/sistema_producao/spcebola/pragas.htm >. Accessed September 13, 2017.

RESENDE, G. M.; COSTA, N. D.; SOUZA, R. J. Onion cultivation. Embrapa

Semiàrido, Production System. 2016.

SEAB, 2016. State Secretariat for Agriculture and Supply. Olericultura - Anâlise da Conjuntura Agropecuària (February 2016). Available at: <http://www.agricultura.pr.gov.br/arquivos/File/deral/Prognosticos/2016/Olericultura_2015_16.pdf>. Accessed on: 06 Nov. 2017.

SHEPHERD, A. W. A guide marketing costs and how to calculate them. FAO, 23 p. 1993.

SILVA, N.F.; HEREDIA, M.C.V.; Consortium of vegetables. *Seminàrios de Olericultura*, v. 7, p.1-19. 1983.

SIMOES, R.; BARRETO, L.; SANTOS, C. A.; SANTOS, J. C.; QUEIJADA, E.; BITTENCOURT, C.; COSTA, N. Cheiro-verde: saber como cultivar hortaliças para semear bons negócios. *Brasilia: SEBRAE*, p. 16-19. 2011.

SOBREIRA FILHO, M. G. *Manual de cultivo das hortícolas em horta domèstica, educativa e comunità.* p. 84. 2012.

THEUNIVERSITY OF MAINE. Available at   : <https://extension.umaine.edu/ipm/ipddl/plant-disease-images/allium-rust-pathogen-images/>. Accessed on: September 20, 2017.

TRANI, P.E. Liming and fertilization for vegetables under protected cultivation. 2007.

TRANI, P.E. CARRIJO, O. A. Fertirrigation in Vegetables. Campinas, Agronomic Institute, p. 58. 2004.

VADA, G. Main types of packaging used in the Sâo Paulo wholesale market. Sâo Paulo: CEAGESP. 1999.

VAZ, A. P. A.; JORGE, M. H. A. Chives. *Embrapa Pantanal- Folders/ Booklets (INFOTECA-E)*. 2007.

ZARATE, N. A. H.; VIEIRA, M. C. Production and gross income of single spring onion and spinach. *Horticultura Brasileira*, Brasilia, v.22, n.4, p.811-814. 2004.

I want morebooks!

Buy your books fast and straightforward online - at one of world's fastest growing online book stores! Environmentally sound due to Print-on-Demand technologies.

Buy your books online at
www.morebooks.shop

Kaufen Sie Ihre Bücher schnell und unkompliziert online – auf einer der am schnellsten wachsenden Buchhandelsplattformen weltweit! Dank Print-On-Demand umwelt- und ressourcenschonend produziert.

Bücher schneller online kaufen
www.morebooks.shop

Printed by Books on Demand GmbH, Norderstedt / Germany